The use of micro-organisms in schools

Department of Education and Science

Education Pamphlet Number 61

London: Her Majesty's Stationery Office

ED Slade pamphlet A019122
576 07 DEP

ISBN 0 11 270391 7

Contents

The use of micro-organisms in schools

1. This pamphlet has been written by H M Inspectors for teachers who wish to give their pupils some practical experience of micro-organisms and their effects. It should interest teachers of biology, primary and middle school science, and home economics; those responsible for courses in health education and environmental studies; teacher trainers; and local authority advisers. Nothing said is to be construed as implying Government commitment to the provision of extra resources.

Why teach microbiology?
2. Microbes are a diverse and fascinating group. Methods of movement and of trapping prey among the protozoa can be of absorbing interest. The production and dispersal of spores in the fungi, and the beautiful patterns on the shells of diatoms, for example, exert an intellectual and aesthetic appeal in their own right, and the microbes illustrate dramatically the characteristic biological processes of living organisms. Work of this kind can make an important contribution to a general awareness of dangers to personal and public health in, for example, the unhygienic handling of food, the transmission of *Salmonella* and of venereal disease. The human body has evolved effective defences against many microbes which, together with sensible precautions and competent technique, should ensure the safety of pupils and staff in following such studies.

Kinds of micro-organisms
3. The microbes or micro-organisms are living organisms of microscopic size. They include many algae, the smaller fungi (moulds and yeasts), the protozoa, the bacteria and the viruses.

4. Like all living creatures, the microbes have particular requirements for survival and growth, and they show adjustment or adaptation to a wide range of environments. Some can survive on or in the bodies of larger animals or plants, so forming relationships which may be co-operative and mutually beneficial (symbiotic), neutral (commensal or cohabiting) or harmful to some degree (parasitic). Most are free-living, however, and essential in maintaining the balance of nature.

5. The algae are simple plants with a firm cell wall and pigments such as chlorophylls which permit photosynthesis. Some are motile, but

most are non-motile single cells or collections of independent cells. The nucleus of most of them has a membrane (eukaryotic nucleus); however the blue-green algae have no nuclear membrane (prokaryotic nucleus) and thus show a similarity to the bacteria.

6. The fungi are also plant-like, with chitinous cell walls, but they lack chlorophyll and feed by using enzymes to digest complex organic substances. Many of them are saprophytes and digest the bodies of dead animals or plants; they are thus important agents of decay. Some can survive in living plants and animals, while others are true parasites and are normally found in living organisms. Some cause disease and are termed pathogens.

7. The protozoa are animal-like and some are parasites of man and his livestock, a consideration to be borne in mind when animals are kept or brought into schools.

8. The bacteria are ubiquitous and are usually smaller than the other microbes. Structurally they differ from other groups (except the blue-green algae) in having no membrane around the nucleus of the cell and no distinct pattern of nuclear division such as that found, for example, in the algae and protozoa. They exploit almost every conceivable environment, including such unlikely places as concrete beams, rusting iron, and fuel oil. We are perhaps most aware of those which cause disease in man, his livestock and his crops—Hugh Nicol* refers to these pathogens as 'homicidal maniacs'—but most bacteria are not merely beneficial, they are essential in making available simple materials which can be absorbed by plants, so completing the circulation of carbon and nitrogen in nature.

9. The viruses are the smallest of all, consisting of nucleic acid surrounded by a protein coat. They are capable of passing through bacteria-proof filters and are visible only by means of the electron microscope. They are detected by their effects upon animals or plants, for they cannot flourish outside living cells. They so alter the chemistry of the infected cell that it produces virus material rather than its own. Some viruses, called bacteriophages, infect bacteria.

Recent changes in school courses
10. For many years biology courses in secondary schools have included a study of fungal, algal and protozoan types together with some of the activities of bacteria. More recently the emphasis has been upon their ecology, activities, and relevance to man, and microbiological techniques are being introduced into the early years of the secondary school. It is now commonplace in many schools to find bacteria and moulds cultivated on nutritive jellies, most

* Nicol, H. *Microbes by the million.* Penguin, 1945

frequently as casual wild growths from the environment or, less frequently, as known pure cultures purchased from biological suppliers. Appendix 1 lists the topics and Appendix 2 a selection from the many organisms suggested in the various science teaching projects to date, though, of course, some modern textbooks may suggest other species.

The primary years

11. During the early years of schooling science is not always identified as such by pupils or teacher; more often it forms a part of some broad topic of interest. Science holds fascination for younger children; it can be seen in the activities of the collector who delights in close observation, in discovery of the unsuspected in the familiar, and who searches for some level of understanding derived from interpreting observations, asking questions of the teacher, and reference to books. Such activity may lead naturally to experiment.

12. The aim in these years will be to introduce the children to the life activities of creatures which they cannot see with the naked eye, rather perhaps than to the creatures themselves. The material used therefore will be that which they find around them: mildews and rusts on weeds and garden plants; soil; baker's yeast; pond dippings; mouldy cheese and bread; hay or grass infusions in rain-water. A hand-lens, a simple microscope or, perhaps better at this stage, a stereo-magnifier, will allow the children to get some idea of the individual organisms in the colony, and in some cases to relate the growth of the colony to physical, or chemical, conditions which either prevail or can be contrived. They will be intrigued by the fact that some moulds are thought to increase the palatability of some cheeses, while others most decidedly do not. They may well wish to enquire into how the cheesemaker can ensure that only the moulds he wants are introduced into the cheese. A mould garden, using bread, say, on a bed of moist sand in a large screw-topped jar will show the variety of colonies which can develop and will stimulate discussion and description in words or pictures. Note that disposal of such cultures needs care; see pp 18 and 22 para. 49 (ii).

13. It is neither advisable nor necessary to attempt to pursue things much further in these early stages, where the main educational gains are likely to be a sharpening of the observational faculties, discriminating use of words to describe things and events, and the heightening of the sense of wonder that such tiny creatures can do so much. Above all, perhaps, the children will be left with an urge to learn more.

Later primary and early middle years

14. Over this period, an increasing number of boys and girls become

ready to probe and test their observations, so transforming the study into experimental science. Perhaps some of the possibilities will illustrate the point.

   i With the mould garden:

      a. Some moulds grow vigorously and spread, others are more compact. Is there a difference in growth rate or simply of habit? Does the vigorous fungus grow on, or merely over, the materials? How do we speed up or slow down growth?

      b. Some moulds show a 'banded' effect as they grow. What is the difference between adjacent bands? Are the bands of equal width? Can we suggest reasons for the banding and how could we test our suggestions? Is there any similarity to 'fairy rings' in pasture lands?

   ii When cress seedlings are grown, those in some containers will probably die. Gardeners call this 'damping off'. Can observations suggest possible reasons? Were they overcrowded? Did they drown? What are the most likely causes? Is there anything peculiar about the dead seedlings? Would one of them infect and kill a healthy seedling? How could one be reasonably sure of getting a culture of healthy seedlings? How could one show that the suspected cause of death was in fact the cause?

   iii Not uncommonly primary classes will do some baking and this provides scope for study of yeast and its use in bread-making. So, too, empirical rules for preserving and storing foodstuffs might well be linked to the conditions needed for growth of moulds.

15. Work so far has largely consisted of isolated activities; valuable as experience, and stimulating to the interest, to the development of meaningful vocabulary and, often, providing the opportunity to investigate experimentally. In later years work will aim at more systematic understanding and an increased recognition of the pupils' varying interests, skills and capabilities.

The later middle and secondary years
16. We may identify three major purposes of science in the secondary years. First, there are important basic ideas which bear directly upon all of our lives and which should properly be regarded as essential elements in the science courses provided for all pupils, the long-term aim being to produce a more scientifically literate population. Second, a number of pupils will take jobs based to some degree upon the application of science, and will gain from having a more detailed appreciation of the subject. Some will work in industrial or public service laboratories, while many will work in the food, retail, and catering industries. They need to appreciate that the basic techniques of sterile handling, of culture and control, are

genuine technical methods of working, drawn from practitioners, rather than classroom substitutes for the real thing. Finally, pupils will need to see the relationship of micro-organisms to general biological principles on the one hand, and to the development of reliable methods of scientific experimentation on the other. All pupils can gain from some study of these aspects of the subject to the extent that their ability and inclination permits.

17. This suggests the possibility of identifying a number of basic elements of microbiology which all pupils can profitably learn, and which may then be extended in various ways to achieve some more sophisticated goals. Much of this work will be developed in science lessons as such, but by no means all of it. Teachers of home economics and of physical and health education will not wish to miss the opportunities presented by their subjects, nor, from different points of view, will teachers of history or geography.

18. The basic elements might be:
  that microbes exist;
  that they are a heterogeneous collection of organisms;
  that they are widespread;
  that certain conditions promote growth, while other conditions limit or inhibit growth;
  that they are essential to the balance of nature;
  that some are used by man for his good, while others are harmful, either causing disease or spoiling products valued by man;
  that they can often be detected by the changes they produce;
  that their activities can be contained by applying techniques of safe handling and disposal.

19. These notions can be established in a variety of ways and Table 1 summarises suggestions from which a selection might be made. In addition, extension work is listed which may be useful at later stages of the secondary school course, or for particular individuals or groups who are progressing quickly.

20. Two further points should be made about Table 1. First, reference is made to some of the people who have contributed to our knowledge of micro-organisms. This is done in the belief that teachers wish to show science to be the product of people who choose to devote their life to this pursuit; and also because it may encourage the reading of original papers. Second, Table 1 represents an educational rather than a specialist approach. For those pupils who are likely to pursue academic courses in sciences such an approach has neither the range nor the precision to satisfy completely. Thus while it may provide opportunity to illustrate some more general biological principles it does not penetrate more

Table. Suggested topics for microbiology courses

Basic microbiology	Suggestions	Some extensions
1 Awareness of existence of microbes distinguishing protozoa, bacteria, moulds, and yeasts	Observation of: Infusions of hay and pond water.	Leeuwenhoek.
a. from the changes they produce;	Mould garden.	Different growth rates; growth 'on' or 'over'; competition.
	Pond water.	Use of monocular and binocular microscopes; dark ground illumination.
b. from microscopical examination;	Decay of leaves; rot of compost in gardening; souring of milk.	Notion of succession; rates and conditions of growth.
c. from growth into colonies or mycelia.	Fermentation of fruits.	
	Wood rot. 'Damping off' of cress.	Notion of 'parasite' and disease. Squash preparations. Koch's postulates.
	Microscopical examination of wet mounts and stained preparations.	Normal staining; negative staining; hanging drop preparations.
2 What promotes growth?	Effect of experimental modifications of conditions on, for example, growth as indicated by CO_2 production and increasing turbidity of yeast in sugar.	Tolerance to desiccation. Notion of thermal death point. Haemocytometer count of cells; absorptiometer or nephelometer readings as further indicators of growth.
a. moisture;		
b. warmth;		
c. light;	Character of colony on nutritive jelly and broth.	Microscopic observation and measurement of germinating fungal spores.
d. pH;		
e. nature of substrate;	Growth tubes and fungi.	
f. oxygen.	Colour change in substrate after seeding with pigmented bacterium, eg *Chromobacterium lividum* (purple).	
3 How to inhibit growth	Distinguish destruction from inhibition and reduction in activity.	

Basic microbiology	Suggestions	Some extensions

	Refrigeration.
	Sunlight.
	Additives:
	  sugar;
	  sulphur dioxide;
	  salt;
	  hops;
	  alcohol.
		Deep-freezing; freeze-drying.
		Use of ionizing radiations commercially.
		Commercial additives, eg:
		  benzoic acid;
		  sodium propionate.

	Sterilisation:
	  heat;
	  chemical;
	  radiation;
	  filtration.
		Processes:
		  canning;
		  smoking;
		  sterilisation;
		  pasteurisation; etc.
		
		Pasteur.
		Tyndall.

	Subculture from a common food, eg milk, at intervals, from portions stored hot; refrigerated; and kept warm.	
4 Helpful microbes	Food microbiology.	

	Yeast:
	  bread-making;
	  brewing;
	  wine-making.	

	Lactic acid bacteria:
	  cheese-making;
	  yoghurt;
	  silage production.	

	Acetic acid bacteria: vinegar-making.	

	Decay:
	  carbon cycle;
	  nitrogen cycle.
		Root nodule bacteria; symbiosis; mycorrhiza.
		Mammalian gut symbionts—cellulose and vitamins.
		Lichens.

	Industrial.	

11

Basic microbiology	Suggestions	Some extensions
	Retting flax. Leather preparation. Production of antibiotics (see below).	Production of: amino acids; organic acids; acetone, butanol, and other organic solvents; enzymes; hormones.
	Treatment of sewage.	
	Filtration. Activated sludge.	
5 Harmful microbes i Illness and disease	Information on venereal diseases.	Ehrlich; Wasserman.
a. causes;	Culture of normal saprophytes from a variety of sources—*care in handling and disposal.*	Semmelweis; Koch; Pasteur.
b. natural body defences;	Bacteriostatic secretions; phagocytosis.	Metchnikoff.
	Immunity and vaccination.	Jenner.

c. therapy:
  disinfectants;
  antiseptics,
  antibiotics.
	Action of disinfectants.	Lister.

	Competition; *Penicillium notatum* and sensitive bacterial species. Use of sensitivity discs.	Fleming. Antibiotic resistance. Public Health controls.
ii Microbial corrosion	Winogradsky column.	Action on concrete and metals, etc.

specialist aspects, for example of genetics and metabolic bio-chemistry, or physiology. At a later stage there could be added:

i growth and population studies, the use of dilution plating techniques and the notion of sampling error;

ii growth requirements, the use of micro-organisms in bio-assay, the influence of the environment on habit and characteristics (eg, fruiting media for fungi and the relationship of genetic inheritance to the substrate requirement);

iii artificial selection (eg, resistance of bacteria to antibiotics, such as in 'hospital disease'); the techniques of isolation from the environment, and enrichment culture;

iv the use of, and limitations of, arbitrary units (eg, the phenol coefficient as a measure of disinfectant activity*);

v study of an environment and the use of culture methods as an isolation technique with inherent errors (eg, inoculation at ambient temperature which inhibits the growth of thermophilic species);

vi particular topics of interest (eg, the viruses and the bacterio-phage; the soil community as illustrated by buried cellulose film which occasionally shows mould predaceous upon nematodes; the problems of timing and dispersal of spores, well exemplified by fungi such as *Coprinus* in animal dung; the phenomenon of bioluminescence, and the opportunity to investigate the effects of heavy metals, and other enzyme-inhibitors, on a suitable culture).

vii Teachers will wish to help their pupils understand how the recently developed techniques of genetic manipulation hold out the possibility of genetic engineering. An authoritative discussion of the factors involved will be found in Cmd. 5880, *Report of the working party on experimental manipulation of the genetic composition of micro-organisms*: HMSO, 1975. No attempt should be made to practise these techniques in schools because the properties of the new organisms thus created cannot as yet be predicted and could prove extremely dangerous. Care should be taken to avoid circumstances in which inadvertent creation of new organisms could occur.

Safety in handling micro-organisms
21. In this, as in other aspects of his work, the science teacher will have regard to the Health and Safety at Work etc Act†, and its requirement that everything 'reasonably practicable' should be done to ensure that hazards are reduced to a minimum.

* Maurer, I.M., Glassware and magic pots. *Laboratory News* 1972, **41**, 16–24
† Health and Safety at Work Etc Act. HMSO 1974

22. A national working party on the laboratory use of dangerous pathogens has published a report* giving guidance on categories of pathogens which present the highest level of hazard, and a Dangerous Pathogens Advisory Group has been appointed. Reference should be made to Administrative Memorandum 6/76 (DES) (4/76, Welsh Office) which shows at Appendix 1 those pathogens which are currently defined as Category A. None of these organisms should ever be used in schools nor should it be necessary ever to use any organism which, although not as virulent as those in Category A, is known to present very considerable hazards which cannot be effectively controlled in school conditions. This pamphlet lists at Appendix 2 the species of organisms suggested in recent science teaching projects which are considered to offer minimal and acceptable risk and so are suitable for school courses.

23. The vast majority of school work can be covered using the acceptable species listed in Appendix 2. If teachers wish to use others, professional advice should first be sought, for example, from the Microbiology in Schools Advisory Committee (MISAC)†, or from the Dangerous Pathogens Advisory Group‡. In each case it is preferable that the channel of communication should, in the case of LEA schools, be through the LEA science adviser. Appendix 3 lists organisms which have been recommended in the past, but which are now known to present unacceptable hazards.

The hazards
24. There are two fundamental reasons why microbes can put a person at risk. First, alien organisms absorbed into the body may find ideal conditions for rapid multiplication, to such an extent that the normal defences against such organisms are overworked and thus permit secondary infection. Second, microbes, like other plants and animals, generate products through their metabolic processes and some of these (the toxins) are harmful, some of them dangerously so. By contrast some toxins are harmful to other micro-organisms; these are the antibiotics which are used to assist in combating infections.

25. Efficient sterilisation, for example of canned meats, will ensure that no live micro-organisms are present, and so will prolonged boiling or cooking of fresh meat or other flesh. If, however, the products were infected prior to sterilisation then toxins may have

---

* *Report of the Working Party on the laboratory use of dangerous pathogens.* HMSO 1975
† Dr. G. Holt, Microbiology in Schools Advisory Cttee. Polytechnic of Central London, New Cavendish St., London
‡ The Secretary, Dangerous Pathogens Advisory Group, Room D410, Alexander Fleming House, Elephant and Castle, London SE1 6JE

been produced which in some cases are thermostable and will not therefore be destroyed by subsequent cooking by whatever means. A common example is the re-heating of pre-cooked food, especially meat or meat-containing foods. The meat itself may be free from infection. When it is cooked and subsequently allowed to cool it may become infected and during both cooling and re-heating provide just the conditions for multiplication of micro-organisms.

26. *Staphylococcus aureus* is commonly present in the nose and on the skin and hair of man. If it penetrates the skin it may give rise to septic pimples and boils. It can cause staphylococcal pneumonia in the lungs, or staphylococcal meningitis when the cerebro-spinal fluid is infected.

27. There are three other sources of hazard. First, bacteria in particular can suddenly mutate, and a few of these changes will be towards increased pathogenicity. Second, individuals vary in their susceptibility to infection: not only may one person be more resistant to infection by a particular organism than is another, but the susceptibility of an individual can vary, often being greater following illness or treatment with certain drugs. Third, a microbe might stimulate an allergic reaction which can be accompanied by unpleasant symptoms.

28. No teacher would knowingly use pure cultures of pathogens. The real danger comes from wild cultures from pupils and environment. Thus, for example, the culture of pupil's finger-dabs is common practice. If one child happens to be incubating a food poisoning organism a near-pure culture may result, with consequent risk if practice is careless.

29. One of the common fungal contaminants of culture plates will be species of *Aspergillus*, a fungus listed among the recommended organisms in science projects (Appendix 2). However, some species are dangerous, eg. *Aspergillus flavus* which is a pathogen and superficially identical with other species. All species of *Aspergillus* should be handled with care.

30. The precautions necessary to ensure that the risk of harm becomes very low indeed are well within the resources of any science laboratory, and require no complicated expertise on the part of the teacher.

Precautions
31. If class discipline cannot be relied on the teacher should abandon individual experiments in favour of teacher-demonstration. It would be easy to say that the first essential is that the teacher should know exactly what it is that is being cultured before the pupils are allowed to handle it, but identification is far from simple and even named,

so-called pure cultures can mutate, and can also very easily become contaminated by unknown and unexpected organisms. Obviously the risk of contamination is much greater if the teacher, technician or pupil preparing the cultures is unwell, and such individuals should not be involved in the preparation of cultures or in dealing in any way with unsealed preparations. **All cultures must in practice be treated as potentially hazardous.** The case of *Serratia marcescens* may make the point. The use of this bacterium, which produces attractive red colonies, has been advocated in curriculum projects in this country and the USA. It was thought to be a harmless saprophyte, though examples of its being an opportunist parasite in man are well known. However, since 1969* some score of cases are known from America in which this micro-organism is associated with grave illness. The species is very adaptable and readily produces strains which are resistant to antibiotics. It is particularly hazardous to the debilitated, the very young, the old, and those undergoing certain medical treatments. **This species should not be used in school.**

32. The problem now becomes one of identifying high risk circumstances and converting them into acceptable minimal hazards. In this analysis the teacher will have regard to his pupils, his technician, his colleagues and himself, and finally to the community outside the laboratory to which contamination may inadvertently be carried.

Good practice

33. Reference should be made to *Safety in science laboratories*, DES Safety Pamphlet No. 2 (HMSO, 1976) and to *Recommended practice in the use of living organisms* (Schools Council, 1974, published by English Universities Press).

34. Source of material and conditions of incubation.

i Where cultures of specific micro-organisms are required they should be obtained only from recognized suppliers.

ii When culturing 'from the wild' it should be borne in mind that pathogens are especially likely to be isolated from the person, from body secretions, from lavatories, and from animal cages.

iii Soil cultures, in addition to carrying pathogenic bacteria such as *Clostridium tetani*, the agent of tetanus, also carry the risk of isolating fungal species, whose spores may cause disease or allergic reactions in sensitised pupils.

iv Incubation at high temperatures isolates thermophilic fungi, some of which produce infective spores.

v Anaerobic culture is best avoided in elementary courses because

* Whalen, T.A. (letter) *Science* 1970, 168, 64

16

of the danger of isolating anaerobic pathogens. However, study of pure cultures of anaerobic sulphate-reducers or photosynthetic organisms at sixth-form level may well be appropriate.

vi Incubation at 37°C tends to select organisms adapted to man's body temperature. Most saprophytic organisms recommended for school microbiology grow well at ambient room temperature.

35. Inoculation.

i A great deal of the microbiology done in schools does not require the use of elaborate equipment, but the use of a transfer cabinet reduces the likelihood of a plate becoming contaminated and helps to reduce general contamination of the laboratory. Cabinets are available from laboratory suppliers, but some schools may prefer to make their own in which case they should be made to a standard not lower than that of a good commercial product. Cabinets should be sterilised after use by exposure overnight to the vapour of 40 per cent formaldehyde*, unless, as is the case with some commercial cabinets, there is provision for ultra-violet sterilisation.

ii G Holt† points out the danger in producing AEROSOLS. Flame-heating an inoculating loop to sterilise it can cause 'spluttering' and aerosol production. The remedy is to dip the contaminated loop into 70 per cent alcohol, drain the loop against the side of the container, and then flame.

iii Quantitative methods, for example dilution-plating, are sometimes necessary. Mouth-pipettes should not be used; disposable syringes often are adequate or, if greater accuracy is needed, rubber bulbs for use with sterilised graduated pipettes are available from most laboratory suppliers.

36. Incubation.

i After inoculation, the top and base of the culture dish should be taped together, to reduce the chance of accidental spillage. Cultures should be labelled on the base with wax pencil to show the nature of the inoculum and the date.

ii Cultures in petri-dishes should be incubated in an inverted position, the base uppermost, so that any condensation will be into the lid rather than upon growing colonies. Care is needed in disposing of condensed liquid to avoid aerosol formation when pouring it into hypochlorite disinfectant.

iii Sterile plates in which heavy condensation has occurred should

* See para 93. DES Safety Series No. 2 *Safety in science laboratories* (revised Spring 1976) HMSO
† Holt, G. Practical tips for the safe handling of micro-organisms in schools. *School Science Review*, December 1974, 56, 195, 248–252

be separated, inverted so as to exclude dust, and dried for a few hours in a cupboard or incubator.

iv Incubation should be at ambient rather than body temperature.

37. Class examination of cultured plates.

Colonies should first be killed by inserting a swab of cotton wool or filter paper soaked with fresh 40 per cent formaldehyde solution into the inverted lid and leaving it overnight. The swab is removed and the base and lip taped together again for class use.

38. Preparation of colonies for microscopic examination.

i Suspensions of a colony in water should be prepared by the teacher and the organisms killed by addition of a drop of formaldehyde. Samples may then be teat-pipetted on to slides.

ii Wet-mount preparations of living organisms to show motility under the microscope should be carried out by the teachers.

39. Selection of culture medium.

Blood agar is used in medical pathology to isolate haemolytic (blood-cell disintegrating) pathogens, and this medium should not be used in elementary courses. McConkey's bile salt medium selectively grows enteric or gut organisms, and it should not be used until the teacher is satisfied that the pupils have adequate experience in microbial techniques.

40. Disposal.

i The simplest arrangement is to dispose of culture-plates, etc., into a bucket of disinfectant, with a separate pot for pipettes, etc. Phenol solution, or commercial products containing phenol, should not be used because it is corrosive and in any case it will not kill spores. A one per cent solution of hypochlorite (1000 parts per million of available chlorine) freshly made-up is usually satisfactory, provided that free chlorine is still being produced. This can be tested by inserting a strip of starch-iodide paper which turns a deep blue if sufficient free chlorine is present. Material treated in this way should be left to soak at least overnight.

ii The safest and least messy procedure is to use autoclavable plastic bags obtainable from laboratory suppliers. Reclaimable glassware, e.g., McCartney bottles (with the cap loosened) are placed into one bag, and disposable materials, including plastic petri-dishes, into another. The bags are then autoclaved and either placed unopened into the dustbin, or the contents removed and washed, as appropriate. While the domestic type pressure cooker is satisfactory, its small capacity can mean that disposal of materials for a whole class can become a tedious process.

iii Some laboratory sinks have open dilution reservoirs which empty by siphoning. They never empty completely and the residue can serve as an effective culture solution especially when the sinks are unused for a period. It is a sensible precaution to pour hypochlorite into a sink after it has been used for the disposal of microbiology material.

41. Maintenance of stock of pure cultures.

i When the stock culture is to be used in class, the normal procedure is to subculture into two or more culture bottles. One of these is incubated, then stored in a refrigerator, to act as the new stock. The other subculture is then used for immediate class supply.

ii When stock cultures are maintained a subculture should be plated out before starting a new series of experiments and examined for signs of mixed growth, indicating that the stock has become contaminated. If mixed cultures do develop then the stock should be autoclaved and replaced by fresh material.

iii Storage. There are advantages in storing stock cultures at 4°C (refrigerator temperature), but such stock should not be stored in refrigerators used to house foodstuffs.

42. Experiments involving pupils.
Some textbooks suggest experiments which involve painting the skin with *Escherichia coli*; or spraying the nasal passage with a suspension of *Serratia marcescens*, or touching colonies through lavatory paper and subsequently culturing from the finger-tips. **No such experiments which involve deliberate contamination of the pupil should be performed.**

43. General.

i All hand-to-mouth operations should be debarred in any laboratory or preparation room used for culturing microorganisms. Particular attention should be given to preparation rooms and to the possibility of staff taking coffee-breaks or sandwich lunches in the room, with consequent risks. There is a case for a small microbiological preparation space, other than the main preparation room, attached to a suite of laboratories. Such a room would be designed to permit easy cleaning (eg, disinfectant spraying) and to avoid dust traps. It would include storage for stock media, dishes, etc; a smooth non-absorbent bench top for preparation, a transfer chamber*, incubator, small refrigerator, and the usual main services.

ii Laboratory coats for staff and pupils are desirable for any work in laboratories, but the case is perhaps stronger in working with

---

* Some transfer cabinets are equipped with motorised extractor fans whose motors are not spark-free. They should not be used with volatile solvents.

micro-organisms in that contamination of the laboratory coat can be contained in a way that is impossible with everyday clothing, for example by soaking in hypochlorite solution.

iii Teachers, technicians, and pupils should make a point of washing the hands thoroughly with soap and warm water after practical microbiological work. Paper towels are to be preferred to roller or other communal towels, which themselves may harbour a rich flora of micro-organisms.

iv Guidance on the use of eye shields is given in *Safety in science laboratories*, No. 2 in the DES Safety Series: HMSO, 1976 (revised edition).

44. Accidental spillages
Use of liquid inoculum and broth cultures means that spillages of droplets or of gross quantities will almost certainly happen eventually. If operations are carried out over a place-mat of lint moistened (not soaked) with one per cent hypochlorite solution, droplets are absorbed and aerosol formation reduced. Alternatively an absorbent pad can be used and disinfected afterwards either by chemical means or by boiling or autoclaving. If hypochlorite solution is used in this or any other situation as a disinfectant, teachers should ensure that it is not less than one per cent concentration at the time of use. Solutions deteriorate in storage.

45. If a spillage does occur the following steps should be taken:–
i Disinfect the area of spillage before starting to clean up, using disposable plastic gloves at this and subsequent stages and allowing about 15 minutes for disinfection to take place.

ii Collect and disinfect contaminated clothing and equipment before they are sent for laundering or washing. Plastic bags should be available for the purpose.

iii See that anyone who has been splashed is cleaned.

iv Dispose of contaminated cleaning cloths etc.

v But where there are reasonable grounds for supposing that dangerous organisms have been dispersed into the atmosphere of a laboratory or elsewhere, evacuate the area and do not attempt to clean it until specialist advice has been obtained from the Medical Officer, who will have access to the Public Health Laboratory.

Some practical considerations

46. Materials and equipment.
The use of specialised growth media and containers is convenient but relatively expensive. The Microbiology in Schools Advisory

Committee has published names and addresses of suppliers, the main culture collections, and lists of equipment*. However, much elementary work can be done with simple equipment. Essentially, experiments will involve a nutrient medium and a source of micro-organisms. The nutrient medium must supply a utilisable nitrogen source, a respiratory substrate (normally sugar), mineral salts, and trace elements. Most of the salts and trace elements are conveniently supplied as yeast extract or peptone. The pH of the medium can be important: thus fungi usually prefer an acid medium while bacteria often prefer a near neutral medium. Convenient and inexpensive media include milk, proprietary stock cubes, strained infant foods, and yeast or meat extract solution. Fluid media can be solidified into gels by addition of 10 per cent by volume of gelatine (though some species of micro-organisms can render gelatine media into fluids). Alternatively 1·5 per cent agar-agar powder can be used. A range of other simple, inexpensive but effective media for fungi is listed in a pamphlet by Dade and Gunnell†.

47. Containers.
Transparent petri-dishes are most convenient as culture dishes. Glass petri-dishes are time-consuming to clean and re-sterilise. The plastic, disposable type are quick to prepare but relatively expensive; they are easily disposed of by incineration after autoclaving. Small jars with screw caps make good culture jars both for stock cultures and class use. Much can be done with soda-glass test-tubes and non-absorbent cotton wool plugs. Alternatively, aluminium caps to fit over unrimmed standard test-tubes are available; indeed aluminium kitchen foil squeezed over the mouth of the tube may be sufficient for preparations which will be handled only by the teacher.

48. Implements.
i Loops can be made by sticking 24 swg Nichrome wire to metal rods; alternatively, and better, chuck needle holders can be purchased. Glass loop holders for class use are unsatisfactory because flame sterilisation, which should include a centimetre or two of the holder, frequently shatters the joint. A few holders in which the wire is straight rather than bent into a loop should be available for use in stab-culture.

ii Swabs are most conveniently obtained as cotton buds designed for cleaning the eyes and noses of infants and re-sterilised in screw-topped glass jars. They may, however, have been treated

* Bainbridge, B.W. *Microbiology in Schools Advisory Committee* (*MISAC*) J Biol Ed. 1972, **6**, 207–210
† Dade, H.A. and Gunnell, J. *Classwork with fungi.* Commonwealth Mycological Institute. 1969

with anti-bacterial substance and it is usually wiser to use a twist of cotton wool on an orange stick, again after sterilisation.

iii Pasteur pipettes are easily made as required if soda-glass tubes, some 20 cm long and 5–7 mm bore, are plugged each end with non-absorbent cotton wool and heat sterilised either in screw-topped jars or wrapped in aluminium foil. The tube is then heated in the middle, pulled apart to form a long taper and heated again to separate completely. A teat is fitted, the sealed end broken off, and the outside flamed to sterilise it. Milk straws can serve as useful, makeshift, non-sterile substitutes.

NB: **Mouth-pipetting should never be used for work of this kind.**

iv Quantitative pipetting: use either sterile disposable syringes or graduated glass pipettes heat-sterilised in aluminium foil and operated by a pipette bulb.

v Spreading. After adding a drop of fluid inoculum to a plate, a convenient spreader is made by re-flaming the pipette and heating some 3–4 cm from the tip so that the end drops at right angles. Allow to cool, then spread the drop over the plate. Alternatively, L-shaped glass rods can be used, after flaming.

49. Sterilisation.

i Heat.

a. Flame: heat inoculating loops to red heat. Note: beware 'spluttering' and aerosol formation—dip first into 70 per cent alcohol, then flame.

b. Hot air oven: for glassware, e.g. pipettes wrapped in aluminium foil and held in a metal pipette container at 140–180°C for $1\frac{1}{2}$ hours.

c. Boiling water (or steam): for glassware or media. Up to 20 hours to kill the most resistant spores. Better to boil 30 minutes, cool and repeat on three successive days.

d. Pressure cooker (autoclave): one atmosphere excess for **at least 15 minutes**. Note that with large amounts of material sufficient time must be allowed for the contents to heat thoroughly.

ii Chemical: a one per cent hypochlorite solution is used to disinfect areas of spillage and in buckets to receive plates etc. Clothing is best disinfected by autoclaving, but proprietary solutions are on the market and can safely be used for all but the largest spills.

50. Sources of organisms.

Considerations governing cultures from the wild have been outlined on page 16.

51. Further reference.

These notes are not intended to be exhaustive. Reference can be made to books in the bibliography or listed as footnotes. Film-loops and slide-tape units are also available to help pupils in the practice of manipulations. Advice on the use of micro-organisms can be obtained through the local education authority science adviser who has access to the local public health laboratory or the local MISAC adviser.

In-service training courses

52. It is possible to qualify as a graduate biologist or to take a main course in science to train as a teacher without having practised aseptic handling of microbes. Furthermore, as increasing numbers of secondary schools adopt a combined or integrated science course, teachers whose main interests or qualifications lie in the physical sciences may find themselves acting as microbiologists for a few weeks. Some schools for younger pupils may not have a science specialist at all.

53. A variety of needs must be met through in-service training, and observation in schools suggests that short courses on elementary microbiological techniques meet the need of teachers and technicians to know the nature of the hazard involved and how the work can be done safely. Elementary courses should afford an opportunity to practise elementary manipulations under supervision. The programme might well include:–

i The use of micro-organisms to illustrate biological principles and educational practices.

ii The nature of the hazards, and safe practice with school classes.

iii Safeguards for the health of the pupils, the technician and the teacher.

The significance of their state of health and of any medical treatment (for example antibiotics or immuno-suppressive drugs) they may be receiving.

iv Elementary techniques.

The use of implements, sterilisation and preparation of media, pouring plates, sterile transfer.

Handling ampoules of freeze-dried organisms, simple screening for contamination, subculturing and isolation.

Maintenance of cultures, basic media for principal groups, selective enrichment media.

Simple classification, staining methods, fermentation media.

54. Some teachers, either in project work or as an extension of course work may wish to pursue with some of their more advanced pupils

a study of micro-organisms well beyond those that have been the major concern of this pamphlet. For them a short course of the kind described above may and probably will be inadequate, and substantial courses nearer a term than a week in length are likely to be needed.

Appendix 1

Microbiology Topics in Science Teaching Projects

Historical

Chain
Ehrlich
Fleming
Florey
Hooke
Jenner
Koch
Leeuwenhoek
Lister
Metchnikoff
Pasteur
Semmelweiss
Spallanzani

Techniques

Handling and culture
Identification
Microscopy
Selective (enrichment) culture

Habitats

Ubiquity
   in air
   in gravy
   in milk
   in soil
   in water
   on ourselves and utensils
   in food

Disease and health

Antiseptics/disinfectants
Sewage

Housefly
Infectious disease
Tooth decay
Immunity and vaccination
Antibiotics

Microbes at work

Fermentation
Cheese, yoghurt, bread
Preservation
Pasteurisation
Spoilage of milk
Spoilage of wine, beer
Decay of food (etc)

Carbon cycle
Nitrogen cycle

Microbes and biological principles

Digestion and enzymes
Competition
Genes and environment

Growth

Nutrition
Symbiosis
Saprophytes and parasites
Mutation

Meiosis
Reproduction
Interdependence
Inheritance
Winogradsky column

Bioassay
Responses

Appendix 2

A list of selected micro-organisms drawn from recent science teaching projects which present minimum risk given good practice.
[Numbers in parentheses refer to the bibliography]

Bacteria

Acetobacter aceti [37]
Bacillus subtilis [34, 35, 36, 41]
Chromobacterium lividum* [43]
Erwinia carotovora
   (=*E. atroseptica*)
Escherichia coli† [36, 41, 43]
Micrococcus luteus
   (=*Sarcina luteus*) [43]
Pseudomonas fluorescens [36]
Rhizobium leguminosarum [36, 37, 41]
Spirillum serpens
Staphylococcus albus‡ [36, 41]
Streptococcus lactis [43]
Streptomyces griseus
Vibrio natriegens
   (=*Beneckea natriegens*)

Fungi

Agaricus bisporus
Aspergillus nidulans [43]
Aspergillus niger [37]
Botrytis cinerea [43]
Chaetomium globosum [43]
Coprinus lagopus [37]
Fusarium solani [12]
   (=*Rhizoctonia solani*)
Mucor hiemalis [36, 37]
Mucor mucedo
Penicillium chrysogenum [12, 36]
Penicillium notatum [36, 43]
Phycomyces blakesleanus [43]
Phytophthora infestans [37]
Pythium debaryanum [35, 36, 37]
Rhizopus sexualis [37]
Rhizopus stolonifer [37]
Saccharomyces cerevisiae [34, 35, 36, 37, 39]
Saccharomyces ellipsoides
Schizosaccharomyces pombe [37, 41]
Saprolegnia litoralis [36]
Sordaria fimicola

Viruses

Bacteriophage (*T type*) (host *E. coli*)

Algae, Protozoa, Lichens, Slime moulds

Though some protozoa are known to be pathogenic, the species quoted for experimental work in the recent science projects, together with the species of algae, lichens and slime moulds quoted, are acceptable for use in schools.

* This species replaces *Chromobacterium violaceum* and *Serratia marcescens*.
† Some strains have been associated with health hazards. Reputable school suppliers will ensure that acceptable strains are provided.
‡ An alternative is *Staphylococcus epidermidis*.

Appendix 3

Micro-organisms suggested in recent science projects but now considered unsuitable for use in elementary courses with pupils below the age of 16

Chromobacterium violaceum
Clostridium perfringens
Pseudomonas aeruginosa
Pseudomonas solanacearum

Pseudomonas tabacci
Serratia marcescens
Staphylococcus aureus
Xanthomonas phaseoli

Bibliography

General

1 L E Hawker & A H Linton — Micro-organisms: function, form and environment — Edward Arnold — 1971
2 A G & P C Clegg — Man against disease — Heinemann Educational — 1973
3 W Bullock — History of bacteriology — Oxford — 1938
4 A J Salle — Fundamental principles of bacteriology (6th edition) — McGraw-Hill — 1967
5 K M Smith — The biology of viruses — Oxford — 1965
6 J H Burnett — Fundamentals of mycology — Edward Arnold — 1968
7 E C Large — The advance of the fungi — Cape — 1940
8 M E Hale — The biology of lichens — Edward Arnold — 1967
9 M A Sleigh — The biology of protozoa — Edward Arnold — 1973
10 F E Round — The biology of the algae (2nd edition) — Edward Arnold — 1974
11 Department of Education and Science — Safety in science laboratories (2nd edition) — HMSO — 1976

Primary and middle years

12 Elementary Science Study Project USA — Teachers' guide for micro-gardening / The micro-gardening cookbook / Illustrated handbook of some common moulds — McGraw-Hill — 1965

Background

13 Clifford Dobell — Antony van Leeuwenhoek and his little animals — Dover Books — 1960
14 Clyde M Christensen — The moulds and man (3rd edition revised) — Oxford — 1965
15 S J Holmes — Louis Pasteur — Dover Books — 1961
16 Robert Reid — Microbes and men — BBC Publications — 1974
17 John Postgate — Microbes and man — Penguin — 1969
18 John Humphries — Bacteriology — John Murray — 1974

Particular topics

19 J A Boycott — Natural history of infectious disease — Edward Arnold — 1971
20 Sir Macfarlane Burnet & D O White — Natural history of infectious disease (4th edition) — Cambridge — 1972
21 Open University — Science and the rise of technology since 1800: Science and public health, Unit 10 — Open University — 1973
22 Betty C Hobbs — Food poisoning and food hygiene (3rd edition) — Edward Arnold — 1974
23 W C Frazier — Food microbiology (2nd edition) — McGraw-Hill — 1967

24	H R Barnell	Biology and the food industry		Edward Arnold	1974
25	V Cheke	The story of cheese-making in Britain		Routledge & Kegan Paul	1959
26	J G Carr	Biological principles in fermentation		Heinemann Educational	1968
27	D F Gray	Immunology (2nd edition)		Edward Arnold	1970
28	G H Booth	Microbiological corrosion		Mills & Boon	1971
29	Alan Burges	Micro-organisms in the soil		Hutchinson	1958

Additional references

30	M Kendall	Preservation of demonstration place cultures	*Housecraft*	Association of Teachers of Domestic Science	1969	146–50
31	MISAC	Review of practical manuals, of value to school teachers using micro-organisms	*J Biol Ed*	Institute of Biology	1971	**5**, 331–40
32	I M Maurer	Glassware and magic pots	*Laboratory News*	World Media	1972	**41**, 16–24
33	Nuffield Foundation	Science Teaching Project. Junior science Series: Teachers' guide (2)	—	Collins	1967	
	M Hardstaff (ed)	Animals and plants				
		Teachers' background booklets:				
	L A Morgan and R. W. Carlisle	Autumn into Winter Science and history				
34	Schools Council	Science 5/13 Project: Minibeasts: a unit for teachers Stage 1 and 2 Change: a unit for teachers Stage 1 and 2	—	Macdonald Educational	1973	
35	Nuffield Foundation	Combined science project: Teachers' guide 1 Teachers' guide 3	—	Longman/ Penguin	1970	
36	A J Mee, Patricia Boyd and D Ritchie	Science for the 70s: Book 1 Teachers' guide Book 2	—	Heinemann Educational	1971	
37	Nuffield Foundation	O-level Biology Project: Teachers' guides 1–4 Pupil texts 1–4	—	Longman/ Penguin	1966	
38	Nuffield Foundation	A-level Biology Project: Teachers' guide to the laboratory guides	—	Longman/ Penguin	1970	
	J A Barker	Control and co-ordination in organisms				
	A L Brown	Key to pond organisms				
	J H Gray	Organisms and population				
	M S Sands	The developing organism				
	C F Stoneman	Maintenance of the organism Study guide: evidence and deduction in biological science				

39	Schools Council	Schools Council Integrated Science Project (SCISP) Patterns: Teachers' handbook	—	Longman/ Penguin	1973
40	Schools Council	Schools Council Project Technology: Food science and technology	—	Heinemann Educational	1973
41	Schools Council (P J Kelly and J D Wray, ed)	Educational use of living organisms: a source book	—	EUP	1975
42	Nuffield Foundation	Secondary Science Project: Themes 1, 2, 3 and 7	—	Longman	1971
43	American Institute of Biological Sciences	Biological Sciences Curriculum Study: Biology teachers' handbook (2nd edition)	—	Wiley	1968
44	Schools Council (P J Kelly and J D Wray, ed)	Micro-organisms	—	EUP	1975
45	H A Dade and J Gunnell	Class work with fungi	—	Commonwealth Mycological Institute	1969

31

Index

The numbers refer to paragraphs or, where appropriate, appendices.

The following titles are among those published and sold by HMSO on behalf of DES.

Safety Series 1	Safety in Outdoor Pursuits
Safety Series 2	Safety in Science Laboratories
Safety Series 3	Safety in Practical Departments
Safety Series 4	Safety in Physical Education
Safety Series 5	Safety in Further Education
Safety Series 6	Safety at School: General Advice

Further details of these and other DES publications can be obtained from PMIC (SL2), HMSO, Atlantic House, Holborn Viaduct, London EC1P 1BN.

051600

Printed in England for Her Majesty's Stationery Office
by Hobbs the Printers of Southampton
(1051) Dd.0585921 K40 6/78 G3927